W9-CGK-862

BY THE SAME AUTHOR
What to Do About Your Brain-Injured Child

The Gentle Revolution Series:
How to Teach Your Baby to Read

Teach
Your Baby
Math

by

GLENN DOMAN

SIMON AND SCHUSTER | NEW YORK

1 2 3 4 5 6 7 8 9 10

Library of Congress Cataloging in Publication Data

Doman, Glenn J
  Teach your baby math.

  1. Mathematics—Study and teaching (Preschool)
2. Infants.   I. Title
QA135.5.D65   572.7   79-17163
ISBN 0-671-25128-7

This book is dedicated with understanding to all of us who ever wondered *why* you put down the two and carried the seven.

It is offered in comradeship to all of us who had arithmetic teachers who were bigger than we were.

It is offered in total empathy with all of us who didn't really *like* math in school, who still don't really *understand* math, and who are still not very confident in adding up the grocery list.

It is offered in brotherhood to all of us who were ever puzzled as to how four dollars' worth of plastic called a calculator can do things that we—with our incredible brains—have not been able to do.

In short—this book is dedicated to almost everybody alive—who is more than two years old.

With a little bit of luck—and some attention to our babies—we'll be the last of the lot.

Contents

Introduction

Dear Parents,

Very few people buy a book for the purpose of disagreeing with it.

The fact that you've bought this book means that, no matter how improbable the title sounds, you've got a healthy suspicion that it is possible to teach your baby how to do math, and in that suspicion you are entirely correct.

Indeed you can, and with a degree of success that even you as parents could not have dreamed to be possible.

It will help you to understand how simply this can be done as well as how incredibly far you can take your baby in math, and the great joy that both you and your baby will know in doing it, if you understand the way in which it all came about.

The staff of The Institutes for the Achievement of Human Potential have had a glorious love affair going with mothers for the last thirty-five years. As the director of The Institutes, I must say it has been a great affair, altogether rewarding and fulfilling.

The affair began poorly and was actually forced upon both the parents and us as a sort of blind date. Mutual trust was low and suspicion was high. It would never have happened in the first place if it hadn't been for the hurt kids and their staggering needs. It was their need that forced parents and us into each other's arms.

In the 1940s the parents of severely brain-injured children had no reason to be grateful to professional people and little reason to trust them. In those days the professional people believed that merely to talk of making a brain-injured child well was not only the worst kind of foolishness but that to do so, even as an objective, was somehow deeply immoral. Many professional people still so believe.

We, as professional people who were daily confronted with children who were paralyzed, speechless, blind, deaf, incontinent, and who were universally considered to be hopelessly "mentally retarded," harbored deep suspicion of parents. Even our own early group that was to become the staff of The Institutes for the Achievement of Human Potential began with the unspoken but common professional belief that "all mothers are idiots and that they have no truth in them." This myth, which is still prevalent, has the tragic result that nobody talks to mothers, and the good Lord knows that nobody listens to them.

Beginning with that belief, as we did, it took us several years to learn that mothers, closely followed by fathers, know more about their own children than anybody else alive.

Myths die hard and the process of unlearning is a great deal harder than the process of learning, and for some people, unlearning is simply impossible. It is frightening for me to admit that if the staggering needs of the brain-injured children hadn't forced us into daily nose-to-nose contact with their parents, we would never have learned the truly extraordinary love that parents have for their children, the profound depth of appreciation they have for their children's potential abilities, and the seemingly miraculous accomplishments they can make possible for their children when they understand the very practical way in which the human brain works.

Suspicion dies slowly and true love must be earned. Often, necessity is not only the mother of invention but also the basis for the beginning of love and understanding if neither party can afford the luxury of running away.

Since the brain-injured children needed help desperately, we and the parents were forced into each other's arms in a marriage not merely of convenience but of necessity.

If the hurt children were to have any sort of life worth living it quickly became apparent that both we and their parents were going to have to devote every moment of our lives to bringing this about.

And so we did.

Beginning a project in clinical research is like getting on a train about which we know little. It's a venture full of mystery and excitement, for you do not know whether you'll have a compartment to yourself or be going second class, whether the train has a dining car or not, what the trip will cost or whether you will end up where you had hoped to go or in a foreign place you never dreamed of visiting.

When our team members got on this train at the various stations, we were hoping that our destination was better treatment for severely brain-injured children. None of us dreamed that if we achieved this goal we would stay on the train till we reached a place where brain-injured children might even be made superior to unhurt children.

The trip has thus far taken thirty-five years, the accommodation was second class, and the dining car served mostly sandwiches, night after night, often at three in the morning. The tickets cost all we had, some of us did not live long enough to finish the trip—and none of us would have missed it for anything else the world has to offer. It's been a fascinating trip.

The original passenger list included a brain surgeon, a physiatrist (an M.D. who specializes in physical medicine and rehabilitation), a physical therapist, a speech therapist, a psychologist, an educator and a nurse. Now there are more than a hundred of us all told, with many additional kinds of specialists.

The little team was formed originally because

each of us was individually charged with some phase of the treatment of severely brain-injured children—and each of us individually was failing.

If you are going to choose a creative field in which to work, it is difficult to pick one with more room for improvement than one in which failure has been 100 percent and success is nonexistent.

When we began our work together thirty-five years ago *we had never seen or heard of a single brain-injured child who had ever gotten well.*

The group that formed after our individual failures would today be called a rehabilitation team. In those days so long ago neither of those words was fashionable and we looked upon ourselves as nothing as grand as all that. Perhaps we saw ourselves more pathetically and more clearly as a group who had banded together, much as a convoy does, hoping that we would be stronger together than we had proved to be separately.

We discovered that it mattered very little (except from a research point of view) whether a child had incurred his injury prenatally, at the instant of birth or postnatally. This was rather like being concerned about whether a child had been hit by an automobile before noon, at noon or after noon. What really mattered was which part of his brain had been hurt, how much it had been hurt, and what might be done about it.

We discovered further that it mattered very little whether a child's good brain had been hurt because his parents had incompatible Rh factors, because his mother had an infectious disease such as German measles during the first three months of pregnancy, because there had been an insufficiency of oxygen reaching his brain during the prenatal period, or because he had been born prematurely. The brain can also be hurt as a result of protracted labor, of a fall on the head which causes blood clots on the brain, of a high temperature with encephalitis, of being struck by an automobile, or of a hundred other factors.

Again, while this was significant from the research point of view, it was rather like worrying about whether a particular child had been hit by a car or a hammer. The important thing here was which part of the child's brain was hurt, how much it was hurt, and what we were going to do about it.

In those early days, the world that dealt with brain-injured children held the view that the problems of these children might be solved by treating the symptoms that existed in the ears, eyes, nose, mouth, chest, shoulders, elbows, wrists, fingers, hips, knees, ankles and toes. A large portion of the world still believes this today.

Such an approach did not work then and could not possibly ever work.

Because of this total lack of success, we concluded that if we were to solve the multiple symptoms of the brain-injured child we would have to attack the source of the problem and approach the human brain itself.

While at first this seemed an impossible or at least monumental task, in the years that followed, we and others found both surgical and nonsurgical methods of treating the brain.

First we tackled the problem from a nonsurgical

standpoint. In the years that followed, we became persuaded that if we could not hope to succeed with the dead brain cells, we would have to find ways to reproduce in some manner the neurological growth-patterns of a normal child. This meant understanding how a normal child's brain begins, grows and matures. We studied intently many hundreds of normal newborn babies, infants and children.

As we learned what normal brain growth is and means, we began to find that the simple and long-known basic activities of normal children, such as crawling and creeping, are of the greatest possible importance to the brain. We learned that if such activities are denied to normal children, because of cultural, environmental or social factors, the potential of these children is severely limited. The potential of brain-injured children is even more affected.

As we learned more about ways to reproduce this normal physical pattern of growing up we began to see brain-injured children improve—very, very slightly.

It was about this time, after working for several years with the parents, that our mutual suspicions disappeared. Love and trust were dawning. So thoroughly had we begun to trust our parents' love and innate good sense that we stopped treating the children ourselves and taught the parents *all* we had learned about the brain, laid out programs for the children, and sent the parents home to carry them out. Results got better, rather than declining. Our respect for parents rose considerably.

It was also at about this time that the neurosurgi-

cal components of our team began to prove conclu-
sively that the answer lay in the brain itself, by de-
veloping successful surgical approaches to it.

A single startling method will serve as an example of the many types of successful brain surgery
which are in use today to solve the problems of the
brain-injured child.

There are actually two brains, a right brain and a
left brain. These two brains are divided right down
the middle of the head from front to rear. In well
human beings the right brain (or, if you like, the
right half of the brain) is responsible for controlling
the left side of the body, while the left half of the
brain is responsible for running the right side.

If one half of the brain is hurt to any large degree,
the results are catastrophic. The opposite side of the
body will be paralyzed, and the child will be severely restricted in all functions. Many such children have constant and severe convulsive seizures
that do not respond to any known medication.

It need hardly be said that such children also die.

The ancient cry of those who stood for doing
nothing had been chanted over and over for de-
cades. That cry was that when a brain cell was dead
it was dead and nothing could be done for children
with dead brain cells, so don't try. But by 1955 the
neurosurgical members of our group were performing an almost unbelievable kind of surgery on such
children; it is called hemispherectomy.

Hemispherectomy is precisely what that name
implies—the surgical removal of half the human
brain.

Now we saw children with half a brain in the

head and with the other half, billions of brain cells, in a jar at the hospital—dead and gone. But the children were not dead.

Instead we saw children with only half a brain who walked, talked and went to school like other children. *Several such children were above average, and at least one of them had an I.Q. in the genius area.*

It was now obvious that if one half of a child's brain was seriously hurt, it mattered little how good the other half was as long as the hurt half remained. If, for example, such a child was suffering convulsions caused by the injured left brain, he would be unable to demonstrate his intelligence until that half was removed in order to let the intact right brain take over the entire function without interference.

We had long held that, contrary to popular belief, a child might have ten dead brain cells and we would not even know it. Perhaps, we said, he might have a hundred dead brain cells and we would not be aware of it. Perhaps, we said, even a thousand.

Not in our wildest dreams had we dared to believe that a child might have *billions* of dead brain cells and yet perform almost as well as and sometimes even better than an average child.

Now the reader must join us in a speculation. How long could we look at Johnny, who had half his brain removed, and see him perform as well as Billy, who had an intact brain, without asking the question, What is wrong with *Billy?* Why did not Billy, who had twice as much brain as Johnny, perform twice as well or at least better?

Having seen this happen over and over again, we began to look with new and questioning eyes at average children.

Were average children doing as well as they might?

Here was an important question we had never dreamed of asking.

In the meantime, the nonsurgical elements of the team had acquired a great deal more knowledge of how such children grow and how their brains develop. As our knowledge of normality increased, our simple methods for reproducing that normality in brain-injured children kept pace. By now we were beginning to see a small number of brain-injured children reach normality by the use of the simple nonsurgical methods of treatment which were steadily evolving and improving.

It is not the purpose of this book to detail either the concepts or the methods used to solve the multiple problems of brain-injured children. Other books, already published or at present in manuscript form, deal with the treatment of the brain-injured child. However, that such problems are being solved daily is of significance in understanding the pathway that led to the knowledge that normal children can perform infinitely better than they are doing at present. It is sufficient to say that extremely simple techniques were devised to reproduce in brain-injured children the patterns of normal development.

As an example, when a brain-injured child is unable to move correctly he is simply taken in an orderly progression through the stages of growth which occur in normal children. First he is helped to move his arms and legs, then to crawl, then to creep, then finally to walk. He is physically aided in doing these things in a patterned sequence. He progresses through these ever higher stages in the same manner as a child does in the grades at school and is given unlimited opportunity to utilize these activities.

A program of this kind having been initiated, we soon began to see severely brain-injured children whose performance rivaled that of children who had not suffered a brain injury. And as the techniques improved even more, we began to see brain-injured children emerge who could not only perform as well as average children but, indeed, who could not be distinguished from them.

As our understanding of neurological growth and normality began to assume a really clear pattern, and as our nonsurgical methods for the recapitulation of normality multiplied, *we even began to see some brain-injured children who performed at above-average, or even superior, levels,* without surgery.

It was exciting beyond measure. It was even a little bit frightening. It seemed clear that we had, at the very least, underestimated every child's potential.

This raised a fascinating question. Suppose we looked at three equally performing seven-year-olds: Albert, half of whose brain was in a jar; Billy, who had a perfectly normal brain; and Charley, who had been treated nonsurgically and who now performed in a totally normal way although there were still millions of dead cells in his brain.

What was wrong with nice, average, unhurt Billy?

What was wrong with *well* children?

Although we were by now working seven-day weeks and eighteen-hour days, each day and hour charged with excitement, we were not doing so alone. So also were the parents, whose own excitement came from the unbelievable things that their hurt children were doing. The love affair had, by the early sixties, produced many hurt children who were totally well in every way and several who were superior. They had become so at home. The love affair worked both ways and had now reached a peak from which it would never fall. What indeed was wrong with well children?

For years our work had been charged with the vibrancy that one feels prior to important events and great discoveries. Through the years the all-enveloping fog of mystery which surrounded our brain-injured children had gradually been dissipated. We had also begun to see other facts for which we had not bargained. These were facts about well children. A logical connection had emerged between the brain-injured (and therefore neurologically dysorganized) child and the well (and therefore neurologically organized) child, where earlier there were only disconnected and disassociated facts about well children. This logical sequence, as it emerged, had pointed insistently to a path by which we might markedly change man himself—and for the better. Was the neurological organization displayed by an average child necessarily the end of the path?

Now, with brain-injured children performing as well as, or better than, average children, the possibility that the path extended further could be fully seen.

It had always been assumed that neurological growth and its end product, ability, were a static and irrevocable fact: this child was capable and that child was not.

Nothing could be further from the truth.

The fact is that neurological growth, which we had always considered a static and irrevocable fact, is a dynamic and ever-changing process.

In the severely brain-injured child we see the process of neurological growth totally halted.

In the "retarded" child we see this process considerably slowed. In the average child it takes place at an average rate, and in the superior child, at above-average speed. We had now come to realize that the brain-injured child, the average child and the superior child are not three different kinds of children but rather represent a continuum ranging from the extreme neurological dysorganization that severe brain injury creates, through the more moderate neurological dysorganization caused by mild or moderate brain injury, through the average amount of neurological organization that the average child demonstrates, to the high degree of neurological organization that a superior child invariably demonstrates.

In the severely brain-injured child we had succeeded in restarting this process, which had come to a halt, and in the retarded child we had accelerated it. It had become clear that the process of neurological growth could be speeded as well as delayed.

Having repeatedly brought brain-injured children from neurological dysorganization to neurological organization of an average or even superior level by employing the simple nonsurgical techniques that had been developed, there was every reason to believe that these same techniques could be used to increase the amount of neurological organization demonstrated by average children. One of these techniques is the teaching of very small brain-injured children to read.

Nowhere is the ability to raise neurological organization more clearly demonstrated than when you teach a well baby to read.

By 1963 there were hundreds of severely brain-injured children who could read and read well, with total understanding at two years of age. They had been taught to do so by their parents at home. Some of the parents had also taught their own well tiny children to do so.

We were ready and had all the information we needed to talk to mothers of well children, and so we did.

In May 1963 we wrote an article called "You Can Teach Your Baby to Read" for the *Ladies' Home Journal*. Letters poured in by the hundreds from mothers who had taught their babies to read successfully and who had found great joy in the doing.

In May of 1964 we published a book called *How to Teach Your Baby to Read;* it was subtitled "The Gentle Revolution." It was published in the United States by Random House and in Britain by Jonathan Cape. Today that book is in fifteen languages. The

letters from mothers have continued to come in—
by the thousands—and they still do.

Those letters report three things over and over
again:

1. That it is much easier to teach a one- or two-
year-old child to read than it is to teach a four-year-
old; and easier to teach a four-year-old than to teach
a seven-year-old.

2. That teaching a tiny child to read brings great
happiness to both mother and baby.

3. That when a tiny child learns to read, not only
does his knowledge grow by leaps and bounds, but
so also does his curiosity and alertness—in short,
that he clearly becomes more intelligent.

The mothers also posed exciting new questions
for us to answer, and high among these questions
was, Now that I've taught my two-year-old to read,
shouldn't it be even easier to teach him math, and if
so, how do I go about doing it?

It took us ten long years to answer this question.
At long last we've answered it and taught hundreds
of tiny well kids and hurt kids to do math easily and
with a degree of success that initially left us in
open-mouthed astonishment. Now it is clearly our
job to make that information available to every
mother alive so that each can decide whether or not
she wishes to take the opportunity to teach her own
baby to do math. This book is our way of informing
mothers that it can be done and how to do it.

And so you see, however improbable it sounds,
your suspicion that you can teach babies to do math
has a very firm foundation in fact.

High on the list of things that we ourselves have

learned is that mothers are, by a long shot, the most superb teachers of children this old world has ever seen.

Have a lovely, loving, and exciting time.

Glenn Doman

P.S. There are no chauvinists at The Institutes, either male or female. We love and respect both mothers and fathers, baby boys and baby girls. To solve the maddening problem of referring to all human beings as "persons" or "tiny persons" we have decided in this manuscript to refer to all parents as mothers and all children as boys. Seems fair.

It will be helpful to the reader in understanding that tiny children are learning math if he understands, at least in a sketchy way, how The Institutes operate.

The Institutes for the Achievement of Human Potential are a group of seven Institutes which exist on the same campus in suburban Philadelphia. Three of these Institutes actually deal with children, while the remaining four are all scientific support Institutes or teaching Institutes for professionals or parents.

Of the three dealing with children, the Institute for the Achievement of Physiological Excellence is the oldest, and deals entirely with brain-injured children, designing programs and teaching their parents how to carry these out at home. The Institute for Human Development is for young adults with severe learning problems.

The third, The Evan Thomas Institute, is for teaching new mothers how to teach their babies to read, do math and do a great many other things, and actually developed as a result of what had been done over past years in the other Institutes.

All three of these Institutes have as their objectives raising these infants, children and young adults to physical, intellectual and social excellence.

1

Mothers and Tiny Kids – The World's Most Dynamic Learning Teams

> We mothers are the potters and our children the clay.
>
> —Winifred Sackville Stoner
> *Natural Education*

I begin my day, as do most people, with breakfast, which is pleasant, and my daily dose of depression —the morning paper—which is not. Sometimes, as it recites its litany of horror, of war, of murder, of rape, of cruelty, of insanity, of death and of destruction I put it aside with the feeling that there isn't going to be any tomorrow and if this proves to be the case, then that just might be the best piece of news of all. But the morning paper is one form of reality; happily, it is not the only form. And I have a guaranteed way of putting the world back into instant and delightful perspective.

A hundred yards from my home is The Evan

Thomas Institute with its charming young mothers, its delightful young staff and its joyous and very ordinary, but extraordinary, babies and tiny kids.

I slip quietly into the back of the room, sit on the floor, lean back against the wall, and I watch the world's most important and most gentle revolution taking place. In five minutes my hopes for the world soar, my spirits skyrocket, my perspective is back in place, and once more it's a great day in the morning.

It's a pleasant room, the Japanese room, with its Japanese *tatami* floor, shoji screens and nothing else except extraordinary people and a vibrant feeling of excitement, love and respect so palpable that everyone who enters there can feel it.

On the opposite side of the room and facing me, three staff members in their late twenties are kneeling. Around them in a semicircle and facing them are twenty mothers in their twenties and thirties. Sitting on the floor in front of the mothers are the very ordinary, quite extraordinary, lovely two- and three-year-old kids. Some of the mothers also have a baby in their arms. No one pays the slightest attention to me or to the other observers, which include a college professor, two schoolteachers, a writer from Britain, an Australian pediatrician and a new mother.

A beautiful little blond two-year-old girl is reading aloud. So absorbed is she in what she is reading that she sometimes giggles as she reads a phrase that touches her sense of humor.

The humor is lost on me because she is reading in Japanese. Although I often work in Japan with Japanese children, my small store of Japanese is not up to her reading. When she reads the phrase that makes her giggle, the other children laugh too. She is reading the Japanese, not in English characters but in the ancient Kanji, the language of Japanese scholars.

There is only one Japanese person in the room. Beautiful, kimono-clad Miki Nakayachi, the Japanese *sensei* interrupts to ask the girl a question. Miki's question and Lindley's answer are both in Japanese, so I understand neither. I remind myself to ask Miki what they were saying that so interested everyone.

Lindley finishes, and Janet Doman, the director of The Institute, asks in English, "Who would like to compose some funny Japanese sentences?"

Several hands shoot up, and Janet chooses Mark, who is three years old. Mark bounces up to take his place beside Suzie Aisen, The Institute's vice director. Suzie places several stacks of large cards before Mark. Each contains a single and, to me, undecipherable Kanji ideograph. Some are nouns or verbs, others articles, adjectives or adverbs.

Mark chooses several cards, lays them out on the floor in an order he chooses and reads them aloud. Everyone laughs, and Janet translates, much to my relief. He has written: "The moose sits on the apple pie."

A two-year-old composes the sentence "The elephant is brushing the strawberry's teeth."

And so in an instant thirty delightful minutes pass.

The staff rises and faces the mothers and children. The children stand up with obvious reluctance, and so do the mothers. Gracefully they all bow to each other. It's such a lovely sight that tears come to my eyes, and I look carefully down at my watch to hide them. I hear laughter because a fifteen-month-old boy has bowed so low that he has lost his balance. He laughs too as he picks himself up.

The reluctance to leave the Japanese language, reading and composing class ends as the cavalcade of mothers and extraordinary but ordinary tiny kids troops down the hall toward the next class, which is advanced math.

I remember how astonishingly far we have come in the fifteen years that have vanished so quickly since May of 1963, when the gentle revolution had begun so quietly with the publication of *How to Teach Your Baby to Read.*

When mothers discovered that they could not only teach their babies to read, but could teach them better and easier at two years of age than the school system was doing at seven, they got the bit firmly in their teeth—and a new and almost indescribably delightful world opened up. A world of mothers and kids. It has within it the potential to change the larger world, in a very short time and almost infinitely for the better.

By 1975 a handful of young, bright and eager mothers had discovered The Evan Thomas Institute and The Evan Thomas Institute had discovered them. Together they taught their babies to read, superbly in English and adequately in two or three other languages. They taught the kids to do math at a rate that left them agog, in shocked but delighted disbelief. They taught their one-, two- and three-year-olds to absorb encyclopedic knowledge of birds, flowers, insects, trees, Presidents, flags, nations, geography and a host of other things. They taught them to do Olympic routines on balance beams, to swim and to play the violin.

In short, they found that they could teach their tiny children absolutely anything they could present to them in an honest and factual way.

Most *interesting* of all, they found that by doing so, they had multiplied their babies' intelligence.

Most *important* of all, they found that doing so was for them and for their babies the most delightful experience they had ever enjoyed together. Their love for each other and, perhaps even more important, their respect for each other multiplied.

The Evan Thomas Institute does not actually teach children at all. It really teaches mothers to teach their children. Here then were these young women, at the prime of life, not at the beginning of the end but rather at the end of the beginning. They were themselves, at twenty-five or at thirty-two, learning to speak Japanese, to read Spanish, to play the violin, to attend concerts, to visit museums, to do gymnastics and a host of other splendid things that most women dream of doing at some dim time in the distant future but that for most people are never realized. That they were doing these things with their own tiny children increased their joy in the doing. Guilt at escaping their children had somehow, and magically, been transformed into pride and a real sense of high purpose for themselves, their children and the contributions they would make to the world.

On a particular morning a year or more ago, when

I had arrived at the math class, Suzie and Janet were presenting math problems to the tiny kids faster than I could assimilate the problems. Their answers were correct—not nearly right but exactly right.

"What," Suzie asked, "is 16 times 19, subtract 151, multiply by 3, add 111, divide by 4 and subtract 51?"

"How far is it from Philadelphia to Chicago?" asked Janet. "And if your car gets 5 miles to the gallon, how many gallons of gas will it take to drive to Chicago?"

"Suppose the car gets 12 miles to the gallon?"

I thought about Giulio Simeone and the day I had asked him what 19 squared was.

"361, but ask me something hard with a big answer."

"Okay," I had responded, searching in my mind for something with a big answer. "How many zeros are there in a sextillion?" Giulio, who was three years old and who likes big numbers, pondered for a few seconds. "21," he announced with a smile. I sat down and wrote out a sextillion. There are 21 zeros in a sextillion.

I had seen such splendid things happen many times before, but they never failed to reastonish me. Nor had it ever failed to restore my soul and my faith that tomorrow would be worth seeing and living.

It had taken us ten years to learn how, but we were finally ready to teach all mothers who wanted to know how to teach their babies to do math. Considering how extraordinarily bright babies are and how easily they learn, it is not surprising that we could teach them. What was incredible was that we had learned how to teach them to do math better than their own parents, who had themselves done the teaching.

How could this be, and how had we learned about it?

2

The Long Road
to Understanding

> Man a dunce uncouth,
> errs in age and youth;
> babies know the truth.
>
> —Swinburne

I was stunned.

Could it possibly be as simple as it seemed to be? If it was, how could I possibly have been so abysmally stupid as to miss it when I had been staring at the answer for so many years? If it was true, I had been a damned fool. I hoped that I had indeed been a damned fool.

It was an odd place to have stumbled on the obvious answer and, at least for me, an even odder time. I was in the Okura Hotel in Tokyo, and it was a little after 6 A.M. I seldom wake so early, since I seldom manage to get to bed much before 2 or 3 A.M.

I had gone to sleep a few hours earlier with the problem very much on my mind.

The team and I were in Tokyo, where we go at least twice a year to teach the parents of Japanese children how to multiply the intelligence of their well babies and their hurt kids. We were quite experienced at this, since we also did it twice a year in Britain, Ireland, Italy, Australia and Brazil, just as we did the same thing full time in America.

The Japanese parents, like the other parents we had been teaching at home in Philadelphia and abroad, were succeeding beautifully.

Virtually all the children could read at far younger ages than did average children; virtually all the children had stored thousands of bits of encyclopedic information in their brains on a myriad of subjects. They also did math at speeds that surpassed that of adults, a fact that was at once marvelous and yet somehow distressing to the adults (although it bothered the children not at all since they didn't know the grown-ups couldn't do it).

The class on How to Teach Your Baby Math had been a review class, since virtually all the two- and three-year-olds were already doing it successfully. The parents, who were delighted that they had successfully taught their kids, were extremely attentive, but were still not clear on my explanation of why kids could do math *faster* and *better* than they themselves could.

I knew the reason that they didn't really understand it was that I didn't really understand, and it was I who was explaining it. Both they and we knew beyond doubt that it *was* so, because the children were doing math beautifully.

Neither the parents nor I had been really satisfied with my answers as to *why*.

Was it purely and simply the very basic and different way we had developed to introduce them to math? If this was the answer, why had we not yet found a single adult who could master the same simple system?

I had gone to bed unhappy with my own complex answers to their questions. I had come awake a few minutes before six, completely alert, which is unusual for me.

Was it conceivable that the answer could be so simple and straightforward? I had considered and rejected a hundred more-complex answers.

Could it possibly be that we adults had so long used symbols to represent facts that (at least in mathematics) we had learned to perceive only the symbols and were not able to perceive the actual facts? It was clear that children could perceive the facts, because they were virtually all doing so.

I recalled the sound advice of Sherlock Holmes, who had proposed that if you eliminate all the factors that are impossible, whatever solution remains must be the answer no matter how improbable it appears to be.

It was the answer.

It's astonishing that we adults have succeeded in keeping the secret of doing math away from children as long as we have. It's a wonder that the tiny kids with all their brightness—and bright they are —didn't catch on. The only reason some careless adult hasn't spilled the beans to the two-year-olds is that we adults haven't known the secret either. But now it's out.

The most important secret is about the kids themselves. We grown-ups have believed that the older you are, the easier it is to learn, and in some things this is true. But it certainly is *not* true concerning languages.

Languages are made up of facts called words, numbers or notes, depending on which language you're talking about. In the learning of pure facts, children can learn *anything* we can present to them in a factual and honest way. What's more, the younger they are, the easier it is.

Words, as everyone knows, are written symbols that represent specific, factual things, actions or thoughts. Musical notes are written symbols that represent specific, factual sounds, and numerals are written symbols that represent specific, factual numbers of objects.

In reading, music and math *most* adults do better than *most* kids, but in distinguishing the *individual* words, notes or numbers all kids learn quicker and much more easily than all adults *if they are given the opportunity young enough.* It is easier for a five-year-old to learn facts than for a six, for a four-year-old than a five, for a three-year-old than a four, for a two-year-old than a three. And by George it is easier for a one-year-old than for a two—if you're willing to be patient enough to wait until he's two to prove it.

It is now abundantly clear that the younger one learns to do something the better he does it. John Stuart Mill could read Greek when he was three. Eugene Ormandy could play the violin when he was three; so could Mozart. Most of the great mathematicians, such as Bertrand Russell, could do arithmetic as small children.

In the learning of mathematics tiny children actually have a staggering *advantage* over adults. In the reading of words we adults can recognize the symbol *or* the fact without effort. Thus either the written word *refrigerator* or the refrigerator itself can be called to mind instantly and easily. Learning the language of music is a little more difficult for adults than for children. If we adults can read music at all, it is much easier to recognize the written note than it is to be sure of the precise sound it represents. Many of us are tone-deaf and are totally unable to identify the actual sound even though we may be capable of reading the symbol. Very few of us have "perfect pitch" and can always identify the exact sound represented by the note. Tiny children can be taught with very little effort to have very close to perfect pitch.

In mathematics the advantage that tiny children have is staggering. We adults recognize the symbols that are called numerals with great ease from the numeral 1 to the numeral 1,000,000 and beyond without effort. We are not, however, able to recognize the actual number of objects beyond ten or so with any degree of reliability.

Tiny children can actually see and almost instantly identify the actual number of objects as *well* as the numeral *if they are given the opportunity to do so early enough in life and before they are introduced to numerals.*

This gives tiny children a staggering advantage over all adults in learning to do and actually to *understand* what is happening in arithmetic.

It will be helpful to the reader's total ultimate understanding if she or he ponders that deceptively simple, but in no way simplistic, fact for a few short chapters. We had pondered that problem for several long years.

Here are some *facts:*

1. Tiny children *want* to learn math.
2. Tiny children *can* learn math (and the younger the child, the easier it is).
3. Tiny children *should* learn math (because it is an advantage to do math better and more easily).

We've devoted a short chapter to each of these vital points.

3

Tiny Children
<u>Want</u> to Learn Math

> Children and genius have the
> same master organ in common
> —inquisitiveness. Let child-
> hood have its way and as it
> began where genius begins, it
> may find what genius finds.
>
> —Edward G. Bulwer-Lytton

While naturally, no child wants to learn math until
he knows that math exists, all children want to absorb information about everything around them,
and under the proper circumstances math is one of
these things.

Here are the Cardinal Points concerning a tiny
child's wanting to learn and his fantastic ability to
learn:

1. The process of learning begins at birth or
   earlier.
2. All babies have a rage to learn.
3. Little kids would rather learn than eat.
4. Kids would much rather learn than play.

5. Tiny kids believe it is their job to grow up.
6. Little kids want to grow up right now.
7. All kids believe learning is a survival skill.
8. They are right in so believing.
9. Tiny children want to learn about *every-thing* and right now.
10. Math is one of the things worth learning about.

There has never been, in the history of man, an adult scientist who has been half so curious as is any child between the ages of four months and four years. We adults have mistaken this superb curiosity about everything as a lack of ability to concentrate.

We have, of course, observed our children carefully, but we have not always understood what their actions mean. For one thing, many people often use two very different words as if they were the same. The words are *learning* and *educating*.

Learning generally refers to the process that goes on in the one who is acquiring knowledge, while *educating* is often the learning process guided by a teacher or school. Although everyone really knows this, these two processes are frequently thought of as one and the same.

Because of this we sometimes feel that since formal education begins at six years of age, the more important processes of learning also begin at six years of age.

Nothing could be further from the truth.

The truth is that a child begins to learn at birth or earlier. By the time he is six years of age and begins his schooling he has already absorbed a fantastic amount of information, fact for fact, perhaps more than he will learn in the rest of his life.

Before a child is six he has learned most of the basic facts about himself and his family. He has learned about his neighbors and his relationships to them, his world and his relationship to it, and a host of other facts that are literally uncountable. Most significantly, he has learned at least one whole language and sometimes more than one. (The chances are very small that he will ever truly master an additional language after he is six.)

All this before he has seen the inside of a classroom.

The process of learning through these early years proceeds at great speed unless we thwart it. If we appreciate and encourage it, the process will take place at a truly unbelievable rate.

A tiny child has, burning within him, a boundless desire to learn.

We can kill this desire entirely only by destroying him completely.

We can come close to quenching it by isolating him. We read occasionally of, say, a thirteen-year-old idiot who is found in an attic chained to a bedpost, presumably because he was an idiot. The reverse is probably the case. It is extremely likely that he is an idiot *because* he was chained to the bedpost. To appreciate this fact we must realize that only psychotic parents would chain any child. A parent chains a child to a bedpost *because* the parent is psychotic, and the result is an idiot child *because* he has been denied virtually all opportunity to learn.

We can *diminish* the child's desire to *learn* by limiting the experiences to which we expose him. Unhappily, we have done this almost universally by drastically underestimating what he can learn.

We can *increase* his learning markedly simply by removing many of the physical restrictions we have placed upon him.

We can *multiply* by many times the knowledge he absorbs if we appreciate his superb capacity for learning and give him unlimited opportunity while simultaneously encouraging him to learn.

Throughout history there have been isolated but numerous cases of people who have actually taught tiny children to learn the most extraordinary things including math, foreign languages, reading, gymnastics and a host of other things by appreciating and encouraging them. In *all* the cases we were able to find, the results of such preplanned home opportunity for children to learn ranged from "excellent" to "astonishing" in producing happy, well-adjusted children with exceptionally high intelligence.

It is very important to bear in mind that these children had *not* been found to have high intelligence first and then been given unusual opportunities to learn, but instead were simply children whose parents decided to expose them to as much information as possible at a very early age.

Once a mother realizes that all tiny children have a rage to learn and have a superb ability to do so, then respect is added to love, and one wonders how she could ever have missed it in the first place.

Look carefully at the eighteen-month-old child and see what he does.

In the first place he drives everybody to distraction.

Why does he? Because he won't stop being curious. He cannot be dissuaded, disciplined or confined out of the desire to learn, no matter how hard we try—and we have certainly tried very hard. He would rather learn than eat or play.

He wants to learn about the lamp and the coffee cup and the electric light socket and the newspaper and everything else in the room—which means that he knocks over the lamp, spills the coffee, puts his finger in the electric light socket and tears up the newspaper. He is learning constantly and, quite naturally, we can't stand it.

From the way he carries on we have concluded that he is hyperactive and unable to pay attention, when the simple truth is that he pays attention to everything. He is superbly alert in every way he can be to learning about the world. He sees, hears, feels, smells and tastes. There is no other way to learn except by these five routes into the brain, and the child uses them all.

He sees the lamp and therefore pulls it down so that he can feel it, hear it, look at it, smell it and taste it. Given the opportunity, he will do all these things to the lamp—and he will do the same to every object in the room. He will not demand to be let out of the room until he has absorbed all he can, through every sense available to him, about every object in the room. He is doing his best to learn, and of course we are doing our best to stop him because his learning process is far too expensive.

We parents have devised several methods of coping with the curiosity of the very young child, and unfortunately, almost all of them are at the expense of the child's learning.

He is aware, if we are not, that learning is for human beings a survival skill. His every instinct tells him so.

Since we are less aware, we have unconsciously devised several methods for the prevention of learning.

The first general method is the give-him-something-to-play-with-that-he-can't-break school of thought. This usually means a nice pink rattle to play with. It may even be a more complicated toy than a rattle, but it's still a toy. Presented with such an object, the child promptly looks at it (which is why toys have bright colors), bangs it to find out if it makes a noise (which is why rattles rattle), feels it (which is why toys don't have sharp edges), tastes it (which is why the paint is nonpoisonous) and even smells it (we have not yet figured out how toys ought to smell, which is why they don't smell at all). This process takes about ninety seconds.

Now that he knows all he wants to know about the toy for the present, the child promptly abandons it and turns his attention to the box in which it came. The child finds the box just as interesting as the toy—which is why we should always buy toys that come in boxes—and learns all about the box. This also takes about ninety seconds. In fact, the child will frequently pay more attention to the box than to the toy itself. Because he is allowed to break the box, he may be able to learn how it is made. This is an advantage he does not have with the toy itself, since we make toys unbreakable, which of course reduces his ability to learn.

The truth of course is that the child never saw the toy as a toy in the first place. He saw both the rattle and the box as being simply new materials from which he had something to learn. The hard and sad truth is that all toys and games are invented by adults to put kids off.

Tiny children never invent either toys or games. They invent tools. Give a child a piece of wood and it immediately becomes a hammer—and he promptly hammers Dad's cherry table. Give a child a clam shell and it instantly becomes a dish.

If you simply watch children you will see dozens of examples of this. Yet, despite all of the evidence that our eyes give us, we too often come to the conclusion that when a child has a short attention span, he just isn't very smart. This deduction insidiously implies that he (like all other children) is not very bright because he is very young. One wonders what our conclusions would be if the two-year-old sat in a corner and quietly played with the rattle for five hours. Probably the parents of such a child would be even more upset—and with good reason.

The second general method of coping with his attempts to learn is the put-him-back-in-the-playpen school of thought.

The only proper thing about the playpen is its name—it is truly a pen. We should at least be honest about such devices and stop saying, "Let's go buy a playpen for the baby." Let's tell the truth and admit that we buy them for ourselves.

Few parents realize what a playpen really costs.

Not only does the playpen restrict the child's ability to learn about the world, which is fairly obvious, but it seriously restricts his neurological growth by limiting his ability to crawl and creep (processes vital to normal growth). This in turn inhibits the development of his vision, manual competence, hand-eye coordination and a host of other things.

We parents have persuaded ourselves that we are buying the playpen to protect the child from hurting himself by chewing on an electric cord or falling down the stairs. Actually, we are penning him up so that *we* do not have to make sure he is safe. In terms of our time, we are being penny-wise and pound-foolish.

The playpen as an implement that prevents learning is unfortunately much more effective than the rattle, because after the child has spent ninety seconds learning about each toy Mother puts in the pen (which is why he will throw each of them out as he finishes learning about it), he is then stuck.

Thus we have succeeded in preventing him from destroying things (one way of learning) by physically confining him. This approach, which puts the child in a physical, emotional and educational vacuum, will not fail so long as we can stand his anguished screams to get out or, assuming that we can stand it, until he's big enough to climb out and renew his search for learning.

Does all the above assume that we are in favor of letting the child break the lamp? Not at all. It assumes only that we have had far too little respect for the small child's desire to learn, despite all the clear indications he gives us that *he wants desper-*

ately to learn everything he can, and as quickly as possible.

We have succeeded in keeping our children carefully isolated from learning in a period of life when the desire to learn is at its peak.

Between birth and four years the ability to absorb information is unparalleled, and the desire to do so is stronger than it will ever be again. Yet during this period we keep the child clean, well fed, safe from the world about him and in a learning vacuum.

It is ironic that when the child is older we will tell him repeatedly how foolish he is for not wanting to learn about astronomy, physics and biology. Learning, we will tell him, is the most important thing in life, and indeed it is.

We have overlooked the other side of the coin.

Learning is the greatest game in life and the most fun. All children are born believing this and will continue to believe this until we convince them that learning is very hard work and unpleasant. Some kids never really learn this lesson and go on through life believing that learning is fun and the only game worth playing. We have a name for such people. We call them geniuses.

We have assumed that children hate to learn essentially because most children have disliked or even despised school. Again we have mistaken schooling for learning. Not all children in school are learning—just as not all children who are learning are doing so in school.

My own experiences in first grade were perhaps typical of what they have been for centuries. In general the teacher told us to sit down, keep quiet, look at her and listen to her while she began a process called teaching, which, she said, would be mutually painful but from which we would learn—or else.

In my own case, that first-grade teacher's prophecy proved to be correct; it was painful, and at least for the first twelve years, I hated every minute of it. I'm sure it was not a unique experience.

In my own case (and I suspect in almost everybody else's) it turned out that the teacher could make me sit down, could make me be quiet, could make me look at her, but could not make me listen and think along with her.

During the rest of that year (and it seemed to me like a hundred years) I found myself in deeper oceans than Cousteau ever visited, on the top of Mount Everest long before Sir Edmund Hillary ever scaled its heights and on the far side of the moon thirty-five years before NASA came into being. I would otherwise have found that century I spent in the first grade a time of crushing boredom interrupted as it was with moments of sheer panic when, during my Jungle Explorations, I dimly heard my teacher calling on "Glenn." It wasn't that I didn't know the answer, it was that I didn't know the question.

I dare dwell on my personal experiences in school only because I believe I was the rule rather than the exception.

Particularly was this so in arithmetic. In first grade we were made to memorize long arithmetic tables such as two times two is four. Being a child, I found this to be dreadfully boring but quite easy. Had I been *two* years old it would have been quite interesting and even easier.

In the second grade it seemed briefly as if things

in arithmetic were looking up. The first day of real multiplication seemed hopeful.

"We're going to multiply 23 by 17," said my teacher. "We put it down this way." Here she wrote on the board:

$$23 \\ \underline{\times\ 17}$$

She had me now and I was interested.

"First," she said, "we multiply 7 times 3. What is that, Bobby?"

"21," said Bobby, who had been made to memorize it.

"Yes," said the teacher. "Now we put down the 1 and carry the 2," suiting the action to the words.

"Why do we do that?" I asked with great interest.

"Do what?" my teacher asked, clearly annoyed.

"Why do we put down the 1 and carry the 2?"

"Because it's the *right* thing to do," said the teacher. "Everybody *else* seems to understand it, so we'll go on."

"Now we multiply our 2 by our 7. How much is that, Eleanor?" asked the teacher.

"14," said Eleanor.

The teacher smiled. Not everybody in the class was as stupid as some small boys.

"Now we add the 2 that we borrowed which makes 16, and we put it down here."

"Why do we do that?" I asked.

She turned to me slowly, letting all the class see her endless patience and how it was tried.

"Now, what do you want to know this time, Glenn?"

"Why do we add the 2 that we borrowed to 14 to make 16?" I asked.

"Because," she said with finality, "it is the proper way to do it."

My curiosity was whetted and the fire started. The fire was out of control.

"Why," I persisted, "don't we *subtract* the two we borrowed or why don't we write it down on the *other* side of the 1?"

"Because," said the teacher, "I'm bigger than you are!"

It was the clearest thing anyone ever said to me in all the years in school.

Now of course the conversation I've just described in such detail never actually took place. It *would* have taken place just as I have described it except that I had always known she was bigger than I was. I wasn't very good in arithmetic, but I wasn't stupid enough to miss the fact that my teacher was bigger than I.

She really believed that the *reason* you put down the 1 and carry the 2 was that it was the proper thing to do, the way to do it, and this was enough reason.

I'm sure she believed this because half a century earlier her teacher had told her that the reason you did it that way was that it was the proper way to do it. My teacher had also known that her teacher was bigger than she was.

That this was the *right* way to do it had never seemed to me to be very persuasive or very logical. It still doesn't.

I suppose this is why I've always been apprehensive about mathematics. I'm downright suspicious of all things that are right because somebody (especially somebody bigger than I) *says* they are right. So many such things have turned out *not* to be right.

Learning *is* fun whether teachers think it is or not, and all tiny children know it.

In summary, babies want to learn about everything, they want to learn about it right now and, having no judgment at all, they want to learn about everything with a fine impartiality.

Part of that *everything* is mathematics, and mathematics is worth learning about.

Strangely, mathematics, which is so difficult for adults to learn, is easier for a one-year-old to learn than anything else.

4

Tiny Children
<u>Can</u> Learn Math

*(and the younger the child,
the easier it is)*

> Feel the dignity of a child. Do
> not feel superior to him, for you
> are not.
>
> —Robert Henri

Virtually everybody loves little kids, but very few adults can honestly be said to respect them. This is because we believe that in every way we are superior to kids. We are taller, heavier, smarter. And, we might well add, a good deal more arrogant.

It is true that we are taller than a little child and heavier, but when it comes to smarter, we should be careful about rushing to a conclusion.

ALL BABIES ARE LINGUISTIC GENIUSES

Linguistic ability is a built-in function of the human brain.

Let's consider the absolutely extraordinary ability that all babies have to learn a language, a miracle beyond measure that we all take totally for granted. Understanding and speaking a language is complex beyond belief and is the single factor that most clearly separates us human beings from the other creatures of the earth.

There are 450,000 words in the English language, and there are 100,000 words in a first-rate vocabulary. Those words can be put together in a virtually limitless number of combinations.

Yet in a normal conversation we encode a message as fast as we can speak. We think in thoughts, and when we speak we often have no idea of how the sentence, paragraph or conversation will end. In short, we encode a message into words, sentences and paragraphs as fast as we can talk. This miracle is not the end. As fast as we can talk by encoding our thoughts into words, that same message is being decoded from words, sentences or paragraphs back into thoughts by the listener.

It is not surprising that we sometimes misunderstand each other; it is breathtaking that we most frequently do understand each other.

Only the human brain is capable of this incredible feat. No computer in existence, nor all the computers in existence hooked together, could carry on a human conversation or even approximate one.

Yet we take it all totally for granted. So complex is human language that only a small proportion of adults ever learn a second language and a very, very small percentage ever learn a foreign tongue perfectly.

ALL BABIES LEARN A FOREIGN LANGUAGE PRIOR TO TWO YEARS OF AGE

This miracle of speech is a built-in function of the human brain.

Any adult foolish enough to get himself into a language-learning contest with *any* average infant would be a fool indeed and would learn that adults are *not* brighter than babies when it comes to the dreadfully complicated business of learning a foreign language.

It must be remembered that all babies learn a foreign language prior to two years of age, speak it fluently by four years of age and speak it perfectly (to their own environment) by six years of age.

We must bear in mind that to a baby born today in Philadelphia, English is a completely foreign language. It is to him no more or no less foreign than French, German, Swahili, Japanese or Portuguese.

And who teaches this baby to perform the miracle of learning this foreign language called English? In our adult arrogance we believe that we do, when, in truth, we actually teach him *Mommy, Daddy* and a few dozen other words. He teaches himself the other tens of thousands of words he will learn, by merely listening to us talk. He does so with the use of his superb human cortex. Only we human beings have such a cortex; only we human beings talk in a contrived, symbolic language, and that unique human speech is a product of the unique human cortex. The human brain gives us the capacity for *language,* and we humans have invented *languages,* hundreds of them.

It is equally true, as everyone knows, that if a child is born into a bilingual household he will speak two languages. If he is born into a trilingual household he will speak three languages—and with no more effort than he spent in learning one, which is no effort at all.

So casually do we accept this unbelievable feat that we give it little or no thought—unless, of course, he does not speak. If, because of brain injury, a child does not speak, then his parents are willing to bring him as far as twelve thousand miles to The Institutes in Philadelphia. Thousands do. Then, and only then, does the size of the miracle become obvious.

Let's compare the performances of an average baby with an adult trying to learn a foreign tongue or even with an adolescent.

Again my own experience may be typical. As a child I wanted very much to learn French. Since in those days every one believed (against all the evidence) that the older you were the easier it was to learn a language, the result was that French was not taught until high school. I was eager to learn, and the fact is that I established some sort of record in my high school. I flunked four consecutive years of French. My record was not in flunking—lots of students flunked. My record was in obstinacy. I was the only one who kept trying for four years. *Nobody* in my class came close to learning to speak French.

I can still remember my teacher, with his fingers on the bridge of his nose and his eyes closed, saying to me, "Mr. Doman, that's one of those awful sen-

tences like 'I seen him when he done it.' " I'm sure
that my French teacher has long since gone to his
reward, and I'm equally sure that his reward is not
having to teach young adults French. Mr. Zimmer-
man can rest easily in his grave because I no longer
say, "I seen him when he done it" in French. Now,
forty years and a dozen trips to France later, I can't
really say much of anything in French, and it isn't
because I don't try.

I was simply too old, in my teens. Yet every average French six-year-old speaks French perfectly
to his own environment. If his family members say
the French equivalent of "I seen him when he done
it," then of course so does he. If his dad is the head
of the French Department at the Sorbonne, then
our six-year-old speaks classical French with fine
grammar and he has not yet seen a teacher or heard
the word *grammar*.

What does all this mean, and what does it have to
do with a tiny child's ability to learn math?

Everything.

IT IS EASIER TO TEACH A ONE-YEAR-OLD
A FOREIGN LANGUAGE
THAN IT IS TO TEACH A SEVEN-YEAR-OLD

Linguistic ability is a built-in function of his
human brain.

As we have just seen.

IT IS EASIER TO TEACH A ONE-YEAR-OLD
TO READ A LANGUAGE
THAN IT IS TO TEACH A SEVEN-YEAR-OLD

This too is a built-in function of his human brain.
Tens of thousands of mothers have taught one-,

two- and three-year-olds to read and read well, while 30 percent of children in all school systems fail to read at all or fail to read at grade level. The Philadelphia school system produces many eighteen-year-old high school graduates who cannot read labels on jars. (This deplorable situation is not confined to Philadelphia.)

They have simply been taught too late.

IT IS EASIER TO TEACH A ONE-YEAR-OLD MATH
THAN IT IS TO TEACH A SEVEN-YEAR-OLD

Mathematical ability is a built-in function of his human brain.

English, French, Italian and all other languages contain tens of thousands of basic symbols called words which are combined in endless intricate relationships of phrases, sentences and paragraphs called grammar, which all well human beings master as babies and children.

Math contains ten basic symbols called 1, 2, 3, 4, 5, 6, 7, 8, 9 and 0.

The astounding question is not why babies and tiny kids can do math faster and more easily than adults, but rather why adults who can deal in a spoken language with ease cannot do math faster and more easily than they can talk.

YOU CAN TEACH A BABY ANYTHING
THAT YOU CAN PRESENT TO HIM
IN AN HONEST AND FACTUAL WAY

It is a built-in function of his human brain.

Babies can be taught facts with the speed of summer lightning, which is in itself a fact that staggers the adult imagination. Most especially is this true if the facts are presented in a precise, discrete and nonambiguous way.

Words, musical notes and numbers are particularly precise, discrete and nonambiguous. This is true whether they are written or sounded. The written word *nose* always means nose, and the spoken sound always means nose. The written musical note middle C always means middle C, and so also does the sound of middle C. The written word *six* always means six, and so does the sound six.

These are facts, and kids learn them a mile a minute. The younger they are, the faster they learn them.

The problem is that we adults divide information into two kinds, the kind we call concrete and the kind we call abstract. Concrete things are those we understand easily; abstract things are those we understand less well.

Then, being adults, we very often insist on teaching children abstractions, which are those things we understand least, while depending on children to learn the precise, discrete, nonambiguous facts by themselves.

In short, we insist on giving tiny kids our opinions rather than the facts. In short, we insist on programming into our kids our own opinions, which very often prove to be wrong. We shall see how serious a mistake this is.

That all tiny kids learn thousands of spoken words before they are three and that thousands of kids can read them as well proves that you can teach a baby anything that you can present to him in an honest and factual way. This includes that very factual and much simpler language called math.

THE ABILITY TO TAKE IN RAW FACTS
IS AN INVERSE FUNCTION OF AGE

It is a built-in function of his human brain.

Myths die slowly indeed, even in the face of overwhelming evidence to the contrary. No myth dies more slowly than the belief that the older you are, the easier it is to learn. The truth is exactly the reverse. The older we are, the more wisdom we acquire, but the younger we are, the easier it is to take in facts and the easier they are to store.

It must by now be obvious to the reader, as it is to all those who know the staff of The Institutes, that we have for all babies' ability to learn and all parents' ability to teach, a respect that borders on reverence. Yet I have never seen a two-year-old wise enough not to fall out of a tenth-story window or to drown himself, given opportunity to do so.

Wisdom, the tiny child does not have; but the ability to take in raw facts—in prodigious amounts —he does have, and the *younger* he is, right down to the early months of life, the *easier* this is.

IT IS EASIER TO TEACH A ONE-YEAR-OLD
ANY SET OF FACTS
THAN IT IS TO TEACH A SEVEN-YEAR-OLD

It is a built-in function of his human brain.

When we speak of a "set of facts" we mean a group of *related* facts. Thus a group of portraits of Presidents of the United States would be a set of facts. Cards each containing the flag of a different nation would be a set of facts, cards each containing a different number of like objects would be a set of facts, and so on.

There are huge advantages in presenting facts to a tiny child in sets; this is discussed in great detail in the forthcoming book *How to Multiply Your Baby's Intelligence.*

That a one-year-old learns sets of facts more quickly than a seven-year-old (and that a seven-year-old learns them more quickly than a thirty-year-old) has been demonstrated thousands of times at The Institutes. Mothers teaching such sets of facts at home find that their children learn them— and retain them longest—in reverse order of age, and that the mother herself learns them the most slowly of all—and forgets them the most quickly. We have also found this to be true with the staff itself, to their mixed chagrin and delight.

With all of the sets of facts presented to the tiny children this fact is the most clear when teaching mathematics.

IF YOU TEACH A TINY CHILD THE FACTS HE WILL INTUIT THE RULES

It is a built-in function of the human brain.

Of all of the unusual things this book has to say, it is possible that this quiet point is the most important. To state it in a slightly different way, if you teach him the *facts* of a body of knowledge he will discover the laws by which it operates. A beautiful example of this exists in the mistakes that tiny children make in grammar. This apparent paradox was pointed out by the brilliant Russian author Kornei Chukovski in his book *From Two to Five* (University of California Press).

A three-year-old looks out a window and says, "Here comes the mailer."

"Who?" we ask.

"The mailer."

We look out the window and see the mailman. We chuckle at the childish mistake and tell the child that he is not called the mailer but the mailman.

We then dismiss the matter. Suppose that instead we asked ourselves the question "Where did the child get the word *mailer?*" Surely no adult taught him the word *mailer*. Then where did he get it? I've been thinking about it for fifteen years, and I am convinced that there is only one possibility. The three-year-old must have reviewed the language to come to the conclusion that there are certain verbs (a word he's never heard) such as *run, hug, kiss, sail* and *paint* and that if you put the sound *er* on the end of them they become nouns (another word he's never heard) and you have *runner, hugger, kisser, sailor, painter,* and so on. That's a whale of an accomplishment. When did you, the reader, last review a language to discover a law? May I suggest when you were three? Still we say, it is a mistake because he is not the mailer, he is the mailman, and so the child is wrong. Wrong word, yes, but right law. The child was quite correct about the law of grammar he had discovered. The problem is that English is irregular and thus, to a degree, a problem. If it were regular the three-year-old would have been right.

The tiny child has a huge ability to discover the laws if we teach him the facts.

It is not possible to discover the facts (concrete) if we are taught only the rules (abstractions).

Let's look at this as it applies to math.

IF YOU TEACH A TINY CHILD
THE FACTS ABOUT MATH
HE WILL DISCOVER THE RULES

This is *not* a built-in function of the human brain, since we human beings have invented mathematics and in some ways taught it imperfectly. Not all human beings have invented math as all human beings *have* invented languages. We have lived among several tribes, such as some in the Xingu territory of Brazil, who do not count at all. Others count only to five.

If you teach a child the facts of mathematics, and the facts upon which mathematics is based are numbers—such as one, two, three, four, five, six—rather than numerals—1, 2, 3, 4, 5, 6, or I, II, III, IV, V, VI —he will discover the rules of mathematics which we call addition, subtraction, multiplication, division, algebra and so on. We shall see precisely how he can do this in the chapter on "How to Teach Your Baby Math."

NOTE: In the following pages the word *number* means the actual quantity or true value, while the word *numeral* means the symbol we use to represent the actual quantity.

We human beings are so very much interested in theories and reasons that we tend to obscure reality. Some portions of this book may be a good example of my own need to understand reasons and to explain them.

To prevent our students and ourselves from losing sight of reality, we use a sign containing the following initials—W. K. I. I. S. B. W. D. I.—and all students are required to write them down on the first day of class.

The following dialogue is extremely common between the staff and parents, professionals and students.

> STUDENT: But how do you *know* you can teach tiny babies math (reading, speaking Japanese, playing the violin, etc.)?
>
> INSTRUCTOR: How did the Wright brothers *know* they could fly?
>
> STUDENT: Well, in the end, I suppose, because they did it.
>
> INSTRUCTOR: That's how we know.

W. K. I. I. S. B. W. D. I. means:

We know it is so because we do it.

Tiny children *are* doing math better and more easily than adults. Hundreds of tiny children are presently doing math and doing it with true understanding of what is happening. Only a minute percentage of adults truly understand what actually happens in math.

5

Tiny Children
<u>Should</u> Learn Math

*(because it is an advantage
to do math more easily and better)*

Mathematics possesses not
only truth but supreme beauty.

—Bertrand Russell

There are two vitally important reasons why tiny
children *should* do math. The first reason is the obvious and less important reason: Doing math is one
of the highest functions of the human brain—of all
creatures on earth, only people can do math.

Doing math is one of the most important functions of life, since daily it is vital to civilized human
living. From childhood to old age we are concerned
with math. The child in school is faced with mathematical problems every day, as are the housewife,
the carpenter, the businessman and the space scientist.

The second reason is even more important. Children should learn to do math at the youngest possible age because of the effect it will have on the physical growth of the brain itself and the product of that physical growth—what we call intelligence.

We have spent thirty-five years searching for understanding of how the human brain grows. There are five points of vital importance which all have to do with how the brain grows.

FUNCTION DETERMINES STRUCTURE

This is an ancient and well-known law of architecture, engineering, medicine and human growth. In human terms this law means that I am what I am because of what I do.

Lumberjacks are hard and muscular because they chop down trees all day. People who lead lives that permit no exercise are soft and not muscular. It is obvious that the biceps grow by use. Weight lifters are a clear illustration of this. If I lift a 25-pound weight daily my biceps will grow. If you lift a 50-pound weight daily your biceps will grow even more. You will then have two advantages over me. You will be able to lift twice as much as I, and secondly, if we are both going to lift 25 additional pounds I will have to double my ability but you will have to increase yours only 50 percent. Concerning the muscles, this is well known and understood. What is not well known and not understood is that this is also true of the brain.

THE BRAIN, LIKE THE BICEPS, GROWS BY USE

The entire back half of the brain is made up of incoming sensory pathways. All these incoming pathways can be divided among the five senses. Everything Albert Einstein or Leonardo da Vinci ever learned in life, everything you or tiny children have ever learned in life entered the brain through these five pathways through which we hear, feel, see, taste and smell.

These five pathways actually grow by use. This is to say that the more messages that pass over the visual pathway, the auditory pathway, the tactile pathway, the gustatory pathway and the olfactory pathway, the larger these pathways will grow and the more easily they will operate. The fewer messages that pass over them, the more slowly they will grow and the less efficiently they will operate. If virtually no messages pass over them, there will be virtually no growth.

We have already mentioned the thirteen-year-old idiot who is found chained to a bedpost in an attic and who is an idiot precisely *because* he has been chained to a bedpost.

When a well baby is born he is born with all these pathways (which we must remember constitute half the brain) intact but immature. It is precisely the light, sound, feeling, smell, and taste impulses passing over these pathways which cause them to grow, mature and become increasingly efficient. Which is exactly why children should read, do math, learn a dozen languages, know great art and exercise as many other sensory skills as possible at the earliest possible ages. Reading, as an example, actually grows the visual pathways. Listening to great music grows the auditory pathways—which incidentally is why parents should talk endlessly to their children. Which brings us to the question of *content*.

The content of the message should be of the highest order. Good language goes into the baby's brain as easily as baby talk, Beethoven goes in as easily as "Pop Goes the Weasel," great art goes in as easily as "Kiddy Cartoons." The possibilities are endless.

The idea of "Reading Readiness" (and all the other "Readinesses") is sheer nonsense. To say that a child is ready to read at five or six is not only nonsense, it is downright dangerous to children. Readiness is *created* in children, and if it isn't created, as it usually is by accident, or as it rarely is, on purpose, it won't come about at all. Witness the child chained to the bedpost.

That the brain grows by use has been known to the neurophysiologists for more than half a century. There have been animal experiments by the hundreds which prove this to be so beyond question. Outstanding among the great scientists in this field have been such geniuses as the Russian Boris Klosovskii and the American David Krech.

For many years Krech and his associates at Berkeley divided newborn rats into two identical groups. One group was raised in an environment of sensory deprivation with little to see, feel, hear, taste or smell. The other group was raised in an environment of sensory enrichment with a great deal to see, feel, hear, taste and smell. He would then test the intelligence of the rats in normal life situations and later sacrifice the rats and measure, weigh and examine their brains microscopically.

Krech's conclusions were that rats that were raised in sensory deprivation had small, undeveloped, stupid brains, while rats that were raised amid sensory enrichment had large, highly developed, highly intelligent brains. Such experiments with such conclusions are myriad.

The front half of the brain is composed of the motor pathways by which we respond to incoming information. They also grow by use. Which is why physical "readiness" as a function of age is also nonsense. Tiny kids can and should swim, do Olympic gymnastics, dance and pursue all other worthwhile physical activities at one and two years of age. They should because they can and because both the body and the brain grow by physical use —as does intelligence.

THE BRAIN IS THE ONLY CONTAINER
THAT HAS THIS CHARACTERISTIC:
THE MORE YOU PUT INTO IT THE MORE IT
WILL HOLD

We have just seen that the brain grows by use and that the more you use it, the better it will function.

But is there some limit to the brain's growth?

For all practical purposes the answer to this question would appear to be that the most advanced human being never came close to being fully used. At a very conservative estimate there are in the human brain never came close to being fully used. At a very conservative estimate there are in the human brain more than ten thousand million neu- these neurons has hundreds or even thousands of interconnections with other neurons. The number of combinations and permutations this makes possible simply boggles the mind of all save some theoretical mathematicians. For human purposes the possibilities can be safely said to be virtually end-less.

It is safe to say that the brain could hold more than we could put into it in many lifetimes—but the more you put into it the better it works.

Math is one of the most useful things you can put into the tiny child's brain.

IF YOU IMPROVE ONE FUNCTION OF THE BRAIN
YOU IMPROVE ALL FUNCTIONS TO SOME DEGREE

There are six functions of the brain which are unique to man. Each of these is a function of the unique human cortex. The first three of these functions are motor in nature:

1. MOBILITY: Only human beings can stand erect on two legs and walk in perfect cross-pattern swinging opposite arms and legs in unison.
2. LANGUAGE: Only humans speak in a contrived, symbolic language that conveys ideas and feelings.
3. MANUAL COMPETENCE: Only human beings can oppose thumb to forefinger and write that symbolic language that we have invented.

These unique motor skills are based on three unique sensory skills:

4. VISION: Only humans can see in such a way as to read that symbolic written language we humans have invented.

5. AUDITORY: Only human beings can hear in such a way as to understand that symbolic spoken language we have invented.

6. TACTILE: Only human beings can feel a complex object and identify it by touch alone.

So vastly interconnected are each of these brain functions that if one could imagine these six functions as each being a cannonball each attached to the other by an iron chain, one could see that it would be impossible to raise one of them very high without pulling the others up to some small or large degree. The child who knows the language of reading finds it easier to learn the language of math.

Conversely, it is not possible to hold one of these functions down without to some degree holding down the others.

The blind child does not run as well as the sighted child.

INTELLIGENCE IS A RESULT OF THINKING

The world has looked at this point exactly in reverse. We have believed thinking to be a result of intelligence. Surely it is true that without intelligence we could not think. But which came first, the chicken or the egg?

We humans are born with the glorious gift of the genes of Homo sapiens. They are the genes of Leonardo, Shakespeare, Einstein, Mozart, Pauling, Russell, Dart, Jefferson and a host of others. But the human brain is not a gift until it is used. We are born with the *potential* brain of all the human greats (and all the scoundrels); intelligence is a result of what we do with it. Intelligence is a result of thinking.

Math is one important way of putting huge stores of information into the brain and is an important way of thinking.

6

How to
Teach Your Baby Math

> "Nina, how many dots can you
> see?"
> "Why *all* of them, Grand-
> mother."
>
> —Three-year-old
>   Nina Pinkett Reilly

Most sets of instructions begin by saying that unless the instructions are followed precisely you will
get bad results. In contrast, about the instructions
that follow, it is safe to say that no matter how
poorly you carry them out, your two-year-old is almost sure to learn more than he would if you hadn't
tried. This is one game you win, to a degree, no
matter how badly you play it! You would have to do
it incredibly badly to produce no result.

Nonetheless, the more cleverly you play the
game of teaching your tiny child to do math, the
quicker and the better your child will learn.

There are some things to remember.

Bear in mind that when we use the word *numeral* we mean the symbols that *represent* quantity or true value—such as 1, 5 or 9. When we use the word *number*—such as one, five or nine— we mean the actual *quantity* of objects themselves.

It is in this difference between true value or quantity and its symbolic representation by the use of symbols to represent actual quantity that tiny children find their advantage over adults.

In mathematics the advantage that tiny children have is staggering. We adults recognize the symbols that are called numerals with great ease, from the numeral 1 to the numeral 1,000,000 and beyond, without effort. We are also able to recognize the *facts* from ●, through ●● , ⚫⚫⚫ , and so on up to about with some degree of reliability. This is an advantage to us because we are able to see at a glance that

$$● \; + \; ●● \; = \; ●●●$$

or that

$$⚫⚫⚫ \; + \; ⚫⚫⚫ \; = \; ⚫⚫⚫⚫⚫⚫$$

However, being able to see the fact exactly from

 up to about  becomes very

unreliable. Beyond twelve only a handful of adults

can see the fact of reliably.

Beyond that we have been taught to rely on symbols entirely.

Tiny children can actually see and almost instantly identify exactly the actual number as well as the numeral.

This gives tiny children their advantage over all adults in learning to do and actually to *understand* what is happening in arithmetic.

These facts are true and they are simple. Perhaps all truths are simple, and for that reason very difficult to see. *There are few disguises harder to penetrate than the all-enveloping cloak of simplicity.*

You can teach your baby to do math even if you aren't very good at doing it yourself. This chapter tells you how. It will be even easier if you have already taught him to read.

The best time to teach your child to do math with little or no trouble is when he is about one or two years old. Beyond two years of age, the teaching of math gets harder every year. If you are willing to go to a little trouble, you can begin when your baby is

18 months old or—if you are very clever—as early as 8 months. There are two *vital* factors:

1. *Your* attitude in teaching him.
2. The *materials* you use.

YOUR ATTITUDE COUNTS HEAVILY

In all history there has never been a more incorrect assumption than that children do not want to learn. Children want desperately to learn, and to learn about everything.

Children begin to learn at or before birth, and they learn intuitively. *Human* thought processes begin at the age of about 3 months in an advanced child; at about 18 months in a slow child.

At birth the thought processes for a well child are instinctive and inevitable. After one year's growth, the thought processes are childlike but human— and still inevitable. In all normal children of *any* age, thinking and learning are inevitable. The one-year-old child believes that learning is necessary, inescapable and the greatest adventure in his life.

Learning *is* the greatest adventure in all of life. Learning *is* desirable, vital, unavoidable and, above all, it is life's greatest and most exciting game.

The child continues to believe this and will always believe this—unless older human beings persuade him that it isn't true.

The cardinal rule is that both parent and child must approach math joyously, as the superb game that it is.

The parent must never forget that learning is life's greatest game—it is *not* work. Learning is a dessert—it is *not* a vegetable. Learning is a reward —it is *not* a punishment. Learning is a privilege—it is *not* a denial.

The wise parent will constantly make this clear to the child. The opportunity to play the math game should be given only to children when they are being well behaved. Children who are behaving badly should be denied the opportunity to play.

Never try to fool a child; you won't succeed. If he has misbehaved, the parent must not permit the child to play the math game just because he wants to play it himself. If the child *has* been behaving badly, it will not do for the parent to tell him he has been a good boy and may play the game. The child won't be fooled for an instant. He knows when he has been bad and that you do not get rewarded for being bad and that if he has been bad and is allowed to play the game, math must be a punishment, not a reward.

The length of time you play the game each time must be very short. At first it may be played as often as three times daily, but for only 20 or 30 seconds each time. The parent must stop each session *before* the child wants to stop.

THE MATERIALS

The materials were designed in complete recognition of the fact that math is a brain function. They can be used successfully with nearly all children. The materials are adapted to the capabilities and limitations of the tiny child's visual apparatus as well as of his brain function. They are very carefully designed to meet all his needs, from stages of visual crudeness to visual sophistication, and from brain function to brain learning.

THE MATERIALS ARE

1. One hundred white cardboards 11″ × 11″, each of which contains on one side a number of large red dots (¾″ in diameter). These dots range in number from one dot on the first card to one hundred dots on the last card. On the reverse side of the card there is the numeral that reveals the number of dots to the parent.

2. One hundred paper cards 5½″ × 5″, each containing a red numeral 5″ high, from the numeral 1 on the first card to the numeral 100 on the last card. These will be used to teach the child the numeral *after* he has learned to do addition, subtraction, multiplication, division and equations very, very speedily. He will learn the numerals very quickly and without effort, which is why they are on less permanent material.

You will notice that the materials begin with very large red dots and with very large red numerals. They are so designed in order that the baby's visual pathway, which is initially immature, can distinguish them readily and without effort. Indeed, the very act of doing so will in itself speed the development of his visual pathway so that the size of the dots and numerals can be quickly reduced to normal reading size. Large figures are used because they are clear. They are red simply because red is attractive to the small child.

THE FIRST STEP
(Number Perception or True Value)

Your first step is teaching your child to be able to perceive actual numbers, which are the *true* value of numerals. Numerals, remember, are merely symbols to represent the true value of numbers. You will begin by teaching your baby (at the youngest age possible down to 8 months) the dot cards from one to ten. First you take only the card with one red dot on it (initially do not let the child see any of the materials except the card with the single dot).

Begin at a time of the day when the child is receptive, rested and in a good mood. Use a part of the house that has as few distracting factors as possible —a quiet place where there is nothing else especially attractive for the child to look at. Do *not* have the radio playing—avoid all unnecessary noises. Use a corner of a room that does not contain a great deal of furniture, pictures or whatever else might absorb the child's visual attention.

Now simply hold up the card just beyond his reach and say to him clearly, "This is one." Show it to him *very* briefly, no longer than it takes to say it. Two or three seconds. At first you will be a bit awkward with the cards. As you become able to deal with the cards in a second or less, so will the child perceive a card in less than a second. Then put it face down in your lap.

Give the child no more description and do not elaborate.

Now hold up the second card where he can see it for a second and say, "This is two."

Now proceed to show him each of the cards in turn up to ten. Use as little time as possible for each new card. The entire process should take less than a minute, including getting settled. The more quickly you show them to him, the better you will hold his attention. There is nothing any child loves better than having his mother's undivided and happy attention.

Now play with him and love him for two minutes. Tell him that he's a very clever little boy and how much you love teaching him. Tell him with great enthusiasm.

The first session is now over, and you have spent no more than three minutes altogether including the get-ready time. You should both have had fun.

Do this three times during the day in the same joyous way.

The first day is now over, and you have spent less than ten minutes in the whole process!

Children learn at lightning speed, and if you show him the dots more than three times a day you will bore him. If you show him a single card for longer than a second or so you will lose him. Try an experiment on his Dad. Ask Dad to stare at a card with six dots on it for thirty seconds. You'll find he'll have great difficulty in doing so. Remember that babies perceive much faster than grown-ups.

Bring great excitement to the game each time you play it. Tiny kids have incredible ability to take in facts but have no ability to determine what is of real value and what isn't. He will take all his cues from you. If you approach the math game each time with pleasure and excitement, so will he.

Do *not* test him. Not yet. Babies love to learn, but they hate to be tested, and in that way they are just about like grown-ups. Testing is the opposite of learning. It is full of stress. To teach a child is to give him a delightful gift; to test him is to demand payment—in advance. The more you test him, the slower he will learn and the less he'll want to. The less you test him, the quicker he will learn and the more he'll want to learn. Knowledge is the most precious gift you can give your child. Give it as generously as you give him food. If you are tempted to test your child before he voluntarily demonstrates to you that he has learned faster than you believed possible, resist the temptation.

Do *not* bribe him or reward him with cookies, candy or the like. At the rate he will be learning, you will not be able to afford enough cookies, and he will not be able to eat them and keep his health. Besides, compared with love and respect, cookies are a meager reward for great accomplishment.

Now repeat the procedure in precisely the same joyous way for the next four days. Five days have gone by and he now knows the facts of one to ten cold, but don't make him prove it yet. You have spent a total each day of 3 minutes teaching him and 5 or 6 loving him, and he has made one of the most important discoveries he will ever make in his whole life. He understands how many, which some primitive people never learn in their lives.

THE SECOND STEP
(Expanding Number Perception)

You are now ready to expand your tiny child's ability to perceive numbers far beyond your own capacity or for that matter the ability of any adult except a very, very few.

Now, each day you are going to continue with three sessions, but each new day you are going to add a new number and remove one of the old ones.

Therefore on the sixth day you will omit card#1 from the ten cards in your lap and will add card #11. On the seventh day you will omit #2 and add card #12. Continue in this way until you have exposed your child to the entire group of number cards, which will take just about three months.

If you have given him this knowledge eagerly and joyously and as a pure gift with no demands of repayment on the child's part, he will have already learned something that very few adults in history have ever learned. He will actually be able to *perceive* what you can only *see*. He will actually be able to identify thirty-nine dots from thirty-eight dots or ninety-one dots from ninety-two. He now knows *true* value and not merely symbols and has the basis he needs to truly understand math and not merely to carry out memorized rituals such as "I put down the 6 and carry the 9." He will now be able to recognize at a glance forty-seven dots, forty-seven pennies or forty-seven sheep.

If you have been able to resist testing, he may by

now have demonstrated his ability by accident. In either case, trust him a bit longer. Don't be misled into believing he can't do math this way merely because you've never met an adult who could. Neither could any of them learn English as fast as every well kid does. You will have already begun the third step and the fourth step before you finish the second step.

THE THIRD STEP
(Addition)

By the thirtieth day you will have exposed your tiny daughter or son to dots up to the number of thirty-five. Now as you continue to spend ten minutes a day exposing him to increasing numbers of dots, you begin teaching him addition. It is as simple as teaching him numbers.

Begin by telling him, in the most excited way, that you are going to teach him addition. Do *not* try to explain to him what addition is. Do not elaborate in any way.

Now use the same place and set of circumstances in which you teach him numbers. You have on your lap the cards that contain two, three, four, five, six, seven, eight, nine, and ten. You have them face down with number two on top of the pile. Using the same happy and enthusiastic tone you say to him, "One plus one equals two." You then show him the card with two dots. You show it to him for no more than a second and preferably less. Do not show him one or try to explain what you're doing in any way. Don't try to explain what *plus* or *equals* means. Always use the same words. Say, "One plus two equals" don't say, "One and two comes to." If you teach children the *facts* they will deduce the rules more accurately and more quickly than any adult. It sounds incredible, but it's true. Remember that thousands of mothers have already done so.

Next you say, "One plus two equals three," and you show him the card with three dots. Very briefly. And so you proceed through to "One plus nine equals ten."

The entire procedure should take no more than one minute. If it takes more you will run the risk of losing his attention.

Do this three times on the first day of Addition while you continue to expose him to the dot cards three times a day. It is best to spread these six sessions out evenly over his period of highest receptivity.

On the thirty-first day you are going to teach him two plus two, two plus three and so on up to two plus eight.

On the thirty-second day you are going to teach him three plus two up to three plus seven.

What you are actually teaching him is what the sounds *plus* and *equals* mean because, believe it or not, he already knows addition. If someone says, "Two plus four equals six" to an adult, what he sees in his mind's eye is $2 + 4 = 6$. Because we adults are limited to seeing the symbols rather than the fact.

What the child is seeing is

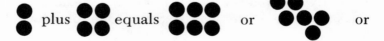

—as all being six. Kids see the fact and not the symbol.

The thirty-third day you will teach him four plus two up to four plus six.

On the thirty-fourth day you will teach him five plus five equals ten.

By the thirty-fifth day he will recognize true numbers up to forty and be able to do addition in every combination up to ten. Most important of all at this stage is that he will completely understand what *plus* and *equals* mean although you will not have told him.

Beginning the thirty-sixth day you should forget about sequence—he has gotten the message. You now give him any addition in any order the sum of which does not exceed the highest number of dots he has been exposed to. Thus on the thirty-sixth day you can begin by saying, "Twelve plus fourteen equals twenty-six." You can follow it by saying, "Seven plus thirty-one equals thirty-eight," and so on. Now give him thirty *new* addition problems every day. Don't do them in any order.

THE FOURTH STEP
(Subtraction)

You will begin to teach him subtraction on the fortieth day. By this time he will recognize dots up to forty-five and will be able to do a huge number of additions up to forty-five.

Again begin by telling him in the most excited way that you are going to teach him subtraction but make no further explanation.

You will now teach him subtraction exactly as you taught him addition. Begin with saying, "Ten minus one equals nine," showing him the nine. Proceed in the same way through "Ten minus nine equals one."

On the forty-first day you may teach him twenty minus nineteen through twenty minus one.

He is now receiving nine very brief sessions a day in this order:

> Numbers Cards
> Addition
> Subtraction
> Numbers Cards
> Addition
> Subtraction
> Numbers Cards
> Addition
> Subtraction

The forty-second day is thirty minus twenty-nine through thirty minus one.

On the forty-third day begin to do subtraction in any form up to forty-eight and in random order.

He now knows numbers up to forty-eight as well as addition and subtraction, and you are spending no more than a total of a half-hour a day, and those thirty minutes, if you're doing it properly, are the greatest and most eagerly awaited minutes of the day.

THE FIFTH STEP
(Solving Problems)

If up to now you have been extraordinarily giving and completely nondemanding, then you are doing very well and you haven't done any testing.

Now you are ready, not to *test* him, but to *teach* him that he knows how to solve problems (and you'll learn that he can).

Begin with numbers. Kneel on the floor facing your tiny kid, who will be sitting on the floor. Take the card with seventeen dots and the card with twenty-five dots and place them both on the floor in front of the child and ask him to point to twenty-five. Ask him casually and happily. Don't feel under the slightest pressure. Both you and he have absolutely nothing to lose and everything to gain. Don't demand that he *say* twenty-five. Tiny kids are still experimenting with language and respond slowly when asked to respond verbally. Remember that you're not trying to teach him to talk—you're teaching him math. He will either look at the correct card, point to it or pick it up. If he doesn't do it almost immediately, then you say cheerfully, "It is this one, isn't it?" as you pick up the card with twenty-five dots. That's true teaching. The chances are good that he'll do it the first time around. If he doesn't, try it again the next day. After a few tries, he will. Whether he identifies it correctly the first time you try or five days later, you must be instantaneously ecstatic. Clap your hands and shout out what a great and brilliant boy he is. Pick him up in your arms and squeeze him. Tell him he's the brightest little boy you ever knew. He is, isn't he?

The moment that you explode with pleasure you will have hooked him on math permanently. He will already have been persuaded that it's more fun than ice cream. Now he'll learn that it's also very rewarding because he has earned the greatest reward life has to offer him—the love and respect of his mother. Test him as little as you're able to do. Remember that knowledge freely given is a golden gift. Testing is a demand for payment.

Now, very occasionally, you can permit him to solve addition problems up to forty-eight. Using the same problem-solving techniques and with your cards arranged in advance, you say to him, "What is seven plus nineteen?" Then you put on the floor two dot cards, one of which contains the correct answer (26 dots) and one which contains a different number of dots. At first use cards with the numbers separated by a half dozen or so dots. Later you can use cards with only a one-dot difference.

Occasionally give him a chance to solve subtraction problems in the same way. Remember to do this infrequently and very quickly.

Do not have specific problem-solving sessions; instead, mix them in by doing one problem occasionally during the teaching sessions. If you do them casually, naturally and without stress, each will add spice to the other.

THE SIXTH STEP
(Multiplication)

You may begin to teach him multiplication on the fiftieth day. By this time he will recognize dots up to fifty-five and will be able to do subtraction and addition up to a total of fifty-five.

Again begin by telling him in the most excited way that you are going to teach him multiplication, but make no further explanation.

You may now teach him multiplication in the exact same way you taught him addition and subtraction. Begin by saying "Two multiplied by two equals four," simultaneously showing him the card with four dots. Now proceed upward to "Two multiplied by five equals ten."

On the fifty-first day begin with "Three multiplied by three equals nine."

By the fifty-eighth day (assuming there have been no badly behaved or grumpy days on either your son's part or yours, when of necessity the math game couldn't be played) you will be up to "Ten multiplied by six equals sixty."

Now in your multiplication sessions you can throw in an occasional opportunity to solve a multiplication problem. You are now having twelve very brief sessions a day with three sessions each in number discrimination or true value, addition, subtraction, and multiplication. Under ideal circumstances your baby has not yet been taught to recognize *any* numeral, not even 1 or 2.

THE SEVENTH STEP
(Division)

You may begin to teach him division on the sixtieth day. By now he knows true value up to sixty-five. While *you* won't be able to tell sixty-one from sixty, you will by now be a superb teacher of your own baby. Perhaps as many as four thousand mothers in the world today have taught their babies to do math, and since there are a little over four billion people alive that makes you just about one in a million. This may be an exciting new thought to you, but your baby has always known that you were one in a million.

Anyway, you are now an expert in how to do it and don't need much instruction. Begin by telling him you are going to teach him division in a joyful way, and then do so. Begin with "Four divided by two equals two" and go to "Sixty-four divided by two equals thirty-two."

By the sixty-eighth day you will have reached "Seventy divided by ten equals seven."

Now throw in an occasional opportunity for him to solve a division problem. You are now doing five subjects each three times daily and are spending a total of about fifty minutes a day, no more.

THE EIGHTH STEP
(Equations)

You may begin to teach him equations on the seventieth day. By now you are a real pro, and we wouldn't dare give you anything except the barest facts. Again begin by telling him you are going to teach him equations and do so in the most excited and happy way. In fact he already knows all two-step equations because, after all, two plus three equals five; seventy minus thirty-one equals thirty-nine; eight multiplied by eight equals sixty-four and seventy-five divided by twenty-five equals three, are all equations. We adults see them only symbolically as $2 + 3 = 5$; $70 - 31 = 39$; $8 \times 8 = 64$

and $75 \div 25 = 3$. Your baby now not only *knows* math but, what is more, actually *understands* what is happening.

Now give him three-step equations. Say, "Seven plus thirteen multiplied by three equals"—then show him the sixty card. After you've taught him many three-step equations move to four-step equations, giving him more-frequent opportunities to solve the equation problems. You will be astonished at the speed at which he solves them. You will wonder if he solves them in some psychic way. When adults see two-year-olds solving math problems faster than adults can, they make the following assumptions in the following order.

1. The child is guessing. (The mathematical odds against this if he is virtually always right are staggering.)

2. The child isn't actually perceiving the dots but instead is actually recognizing the pattern in which they occur. (Nonsense, he'll recognize the number of men standing in a group, and who can keep people in a pattern? Besides, why can't *you* recognize the seventy-five patterns on the seventy-five dot cards he now knows at a glance?)

3. It's some sort of a trick. (You taught him. Did you use any tricks?)

4. The baby is psychic. (Sorry but he isn't: he's just a whiz at learning facts. We'd rather write a book called "How to Make Your Baby Psychic," because that would be even better. Unfortunately, we don't know how.)

Now you can give him five- or six-step equations such as "Seventy-five minus fifty multiplied by two plus ten and divided by three equals twenty," or anything else you like.

By the ninety-fifth day he will know dots up to one hundred and he will be able to do a sort of instant math. Based on an actual understanding of what is *happening*. You will still work on the jingles of symbols you were taught such as "2 times 8 is 16, put down the 6 and carry the 1. 2 times 7 is 14, add 1 you carry makes 15. 3 times 8 is . . ." and so on, ad infinitum, ad nauseam and add lack of true understanding.

Don't bother with teaching him dots beyond one hundred—after that you're only playing with zeros.

Remember that *if you teach a baby the facts he will intuit or deduce the laws.* You can now teach him *any* additions, multiplications, subtractions, divisions or equations up to nine hundred and ninety-nine without bothering to tell him that "one hundred multiplied by two equals two hundred." Good Lord, he already *knows* that.

THE NINTH STEP
(Numerals)

This step is ridiculously easy. Now you take the numeral cards. You hold up the card with the big red 1 on it and simply say, "This is called one." Show him the numeral cards 1 to 25 on the ninety-sixth day. Add 26 to 50 the next day. By the ninety-ninth day he'll know them all.

THE TENTH STEP
(Making the Materials)

Obviously, preparing the materials is the first thing you must do rather than the last. We have placed it last because you will make the cards best after you understand how you are going to use them.

You will need one hundred pieces of good quality, reasonably heavy two-sided white cardboard. (They are going to get heavy use.) These cards should be 11″ × 11″. You can get them from your stationery store, and they'll probably cut them for you to the precise size.

Now you will need 5,050 red dots ¾″ in diameter. You can buy these dots at the stationery store. One variety made by Dennison is called PRES-a-ply. These dots have adhesive backing. You must work very carefully in preparing the cards. The dots must be placed on the cards in a totally random way working outward from the middle. Be certain of three things:

1. That you count out the precise number of dots *before* applying them to the card. (You'll have great trouble in counting them *after* you put them on the card when you get above twenty.)

2. Be sure *not* to place them in a pattern such as a square, circle, triangle or diamond or a shape of any other sort.

3. Be sure to write the numeral on the back of each card *as* you do them.

Now you need only make the 100 paper cards with the numerals on. You can do this with a speedball pen and red ink. The paper cards can be less expensive and thinner than the cardboard cards and should be 5½″ × 5½″ with the numerals 5″ high and

3″ wide up to 9. From 10 to 100 the numerals can be 3″ high and 2″ wide.

Making the materials is a difficult and somewhat expensive job but nothing compared to the thrills and excitement you and your child will have doing math together.

We include here a letter from a mother which states it precisely:

March 22, 1978

To: Glenn Doman
The Institutes, Philadelphia

Re: Math Dot Flash Cards

Dots are terrific! I don't know what I'm doing but the kids love it.

However, I just thought you'd like to know that my cost for the set of 100 cards was

$16.50 (Canadian $s for dots)
15.00 (Bristol Board card @ 10% discount)
10.00 Vinyl covering—⅓rd regular price
_____

which equals 41.50 plus ????? hours of cross-eyed work.

Can't really recommend *that* to my friends!

When are you planning to sell a ready-made set?

P.S. Please hurry with your book. I need help.

Mrs. Laura van Arragon
Ontario, Canada

Note to Mothers:
 You can buy the cards already made up from

The Better Baby Press
The Institutes for the Achievement of
  Human Potential
8801 Stenton Avenue
Philadelphia, Pennsylvania 19118
U.S.A.

However, if you are as excited as we hope you are and can't wait for the mails, start making the materials right now and go to work (play?).
 Now you're ready to begin! Have fun.

Throughout this teaching of math, remember that you should both be looking forward to doing it with great pleasure and anticipation. Learning is one of the greatest joys in life, and it should remain so. Remember, you are building into your child a love of learning that will multiply throughout his life. More accurately, you are reinforcing a built-in rage to learn which will not be denied but which can certainly be twisted into useless or even very negative channels. Play the game joyously. You are giving your child an unparalleled opportunity for knowledge by opening the golden door to all the fascinating problems that can be solved by mathematics.
 Bear in mind that numerals, at which we adults are trained, are abstract and without meaning except as symbols to represent numbers. Actual numbers on the other hand, at which children are su-

perb, are as concrete as it's possible to be. So concrete that the child can actually "see" the number in his mind's eye and can "read" the actual number as we can read only the numeral. That's why the tiny kid can answer the math problem almost the instant you present it.

He's a great deal like those annoying little calculators. We say to our tiny calulator by pushing its buttons, "Calculator, what is 987 multiplied by 654?" We push the *equal* button, and faster than the eye can see, the answer appears—645,498. That's insulting in the extreme. Here is four dollars' worth of plastic and wires answering with a push of a button and without hesitation a problem that we human beings with our vast and incredible human cortex can answer only after much writing. Yet we human beings master English, French or German without the slightest effort when we are children. No computer in the world, no matter how complex, no matter what its multimillion dollar cost, can carry on a true conversation. It's the paradox to end all paradoxes. It is, in short, an absurdity. Happily, the tiny child is quite capable of upholding the honor of the human brain. He, like the four dollars' worth of plastic and wire, is always looking at the answer and knows it instantly. There are in the world a handful of adult mathematical geniuses who can do the same. There is a Dutch Mathematician at the Cern (European Center for Nuclear Research), a man named Willem Klein, who can in two minutes and forty-three seconds solve totally, in his head, the 73rd root of a 499 digit numeral. An electronic computer confirmed the human solution— 6,789,235.

In this respect he is like the tiny children. Who is surprised that he understood math as a tiny child? He does not find it necessary to play the dumb old "carry the 7" rituals by which we were all taught to do math and were thus condemned to a lifetime of painfully complicated (and as a result frequently inaccurate) recipes. It should be mentioned that the author, being already a grown-up, is as condemned to the old methods as the reader is likely to be.

Remember that children are learning every waking minute, and we are teaching them all the time, but the problem is that we are not always aware that we are teaching them and thus may be teaching them something we do not intend to teach them.

Go very quickly or you will make him frantic with boredom. Recently one of our mothers, who was teaching her daughter the dot cards, was going very slowly (because she was afraid that her daughter didn't truly understand the dots) and was very quietly, but very effectively, reprimanded by her three-year-old.

This tiny child in exasperation said, "Oh, Mother!," took the dot cards and selected several of them. Marching into the dining room, she put the card with thirty-two dots on her father's plate, the card with thirty dots on her mother's plate, the card with eight dots on her brother's plate, the card with five dots on her sister's plate and the card with three dots on her own plate. It took her mother a while to figure out that she had placed each person's age in dots on his or her plate. Children who know math ask all sorts of questions about numbers and remember the answers. Her mother got the message and moved five times faster.

Remember that tiny kids don't know that we grown-ups can't do it.

One of our wise grandmothers held up a card with sixty-nine dots and asked her three-year-old granddaughter a dumb question.

"How many dots can you see, dear?"

"Why, *all* of them, Grandmother."

Ask a child a dumb question and you'll get a smart answer.

Historically, we have not only dreadfully underestimated the tiny child's ability to do math, while substantially overestimating our own ability, but we have given him material to work with that falls far short of his ability and his level of enjoyment. It is a miracle we have not bored him to death, instead of just boring him to the point of not wanting to do math. We give five-year-olds problems to solve that would bore a two-year-old to tears. Can you imagine two-year-olds who can do equations in nothing flat faced with a book that explains that two fuzzy bears who meet three fuzzy bears are altogether five fuzzy bears? The idea that tiny children are dehydrated adults who must deal with dehydrated numbers is most certainly a silly one.

Tiny kids can learn *anything* that you can present to them in a factual and honest way. Don't give them theories and abstractions. Give them facts. They'll deduce the laws. Tiny kids operate by the same methods that scientists do.

The dictionary defines science as "that branch of knowledge that deals with a body of facts systematically arranged to reveal the laws."

According to this definition, tiny kids *are* scientists.

You are now an expert teacher. You have taught a small child to do math, something very few people have done. Who are we to tell you how to teach him more? The world of numbers is now his oyster. Using whatever method you think is wise, teach him everything you like. Buy him a small but real calculator after he's off and running. He'll astonish you with what he learns, and the sky's the limit. All the general rules we have proposed still apply.

Do not push him.

Do not bore him. In all cases stop *before* he wants to stop. Do not persist in your teaching for more than a few minutes at a time, but do it often.

Always remember that math is a game. It is fun! It is playing with your baby. It really is. In our experience those mothers who approached the teaching of math with casual gaiety and imagination and who shouted their approval of *every* correct answer with unalloyed enthusiasm, succeeded better than those mothers who showed intellectual objectivity and sober praise. Remember that you are not the Secretary of Education. You're his mother and you're on *his* side. It must *not* be a chore, it must *not* be a grind.

Once your child's free and running, you will have a thinker on your hands and there will be no stopping him. It seems a little silly to say at this point that math, like reading, is basic to all education, to virtually all learning in the world we know. You will have opened the doors to learning, the greatest treasury life has to offer except, perhaps, love and respect (that is, emotional response and responsibility), without which nothing is really worthwhile.

In the process of teaching your baby math, you will both have learned more about love and respect.

Now that the secret is out, it is no longer a question of whether little kids can do math—it's a question of how far they'll take it. The new question will be, we guess—when hundreds of thousands of pre-school kids can do math and thus increase the world's knowledge beyond anybody's wildest dream—what will they do with this old world and how tolerant will they be with this bunch of old parents who, by their new standards, may seem nice but perhaps not very bright?

Afterword

Am I upset by the idea of making little children superior? What's wrong with perfection?

—Walker Buckner

Not all human beings, even bright ones, see instantly what we are proposing for tiny children in "The Gentle Revolution Series," which teaches parents how to teach babies to read, to do math, to gain encyclopedic knowledge, to multiply their intelligence and to improve them physically and socially.

Many are made uneasy or even suspicious that we wish to steal their precious childhood or make them programmed little automatons.

Those concerned feel that tiny children are happier in a learning vacuum, which is simply untrue, and that highly intelligent and knowledgeable people are more likely to be automatons than less-intelligent and less-knowledgeable people. We believe the reverse to be true.

What is it that we *do* want for tiny kids, for parents and for the world?

The world's attitude and ours were summarized neatly a couple of years ago during a television talk show.

We had been talking about what we, through parents, had been teaching tiny kids.

The host was intellectual and bright, but it was obvious that he was becoming increasingly concerned as the conversation progressed. He could stand it no longer.

> HOST (*Accusingly*): But it sounds as if you are proposing some sort of elite!
>
> WE: Precisely.
>
> H: Are you admitting that you propose to create an elite group among children?
>
> W: We are proud of it.
>
> H: Then how many children do you want to have in this elite of yours?
>
> W: About a billion.
>
> H: A billion? How many children *are* there in the world?
>
> W: About a billion.
>
> H: Aha, now I begin to see—but then, who do you want to make them superior *to?*
>
> W: We want to make them superior to *themselves.*
>
> H: Now, I take your point.

And that's what we, the staff of The Institutes for the Achievement of Human Potential, want for ourselves, for your children, for you and for everyone else in the world.

Acknowledgments

The basic ideas that this book contains are so simple
and obvious that it is difficult to believe that someone, or lots of people, haven't seen them before. So
far as we have been able to determine, no one has.
Or at least no one has ever expounded them clearly
enough to be understood or perhaps strongly
enough to be heard.

If, in fact, someone else has expounded them but
has not been heard, I strongly suspect that it was
because he was alone.

Nobody ever wrote a book by himself.

Especially this one.

If *we* are heard it will be because of the following
people:

Janet Doman, my daughter, director of The Evan Thomas Institute, and Suzie Aisen, the vice director. They developed the methods of instruction and did the teaching. They were virtually coauthors.

Gretchen Kerr, the vice director of The Institutes, who so ably took over my duties to add to her own awesome duties while she sent me away to write this book.

Katie Doman, my wife, who first taught mothers how to teach their babies to read, to do math and to multiply their intelligence and who still does it superbly.

Greta Erdtmann, my research assistant; and Cathryn Ruhling, her assistant, who conferred loving care upon the manuscript through the dead of night and upon me as well.

I wish to record my love and devotion to them as well as all the following who make ideas real every minute of every day:

The senior staff of The Evan Thomas Institute: Miki Nakayachi (Japan), Gail Engebretson (United States), Conceicao de Sousa (Brazil), Helen Downey (England), Joan Katz (United States) and Marian Derr (United States).

The incredible mothers and tiny kids of The Evan Thomas Institute.

The director of The Institute for Human Development, my son, Douglas Doman; the vice director, Bruce Hagy; and the senior staff: Holly Purdue (United States), Max Britt (Australia), Margaret Sullivan (United States) and Lidwina van Dyk (Australia).

The director of The School for Human Development, Rosalind Klein; the vice director, Charles Solis; and the staff; Leia Coelho (Brazil), Lorna Rogers (Britain), Anna Maria Massaro (Italy), James Lattanzio (United States) and Mary Jean Poncia (United States).

The courageous students of the School with its supremely courageous and endlessly determined Appalachian Team: Susan Cameron, Helene Milestone, Richard Beebee, Jack Fisher, Etienne Schroeder and John Grifo, heroically led by staff members Charles Solis and Lidwina van Dyk.

The Institute for the Achievement of Physiological Excellence and its director, Mary Kett (Ireland); vice director, Ann Ball (Britain); and its senior staff: Charlotte Coombs (United States), Dawn Price (New Zealand), William Wells (United States), Mitsue Okabayashi (Japan), Hiromi Nakamura (Japan) and Marta Moto (Brazil).

The endlessly determined parents of that Institute and their invincible children.

The Institute for Clinical Investigation and its director, Edward B. LeWinn, M.D.; and its vice director, Pearl LeWinn.

The Institute of Man and its chairman, Professor Raymond A. Dart (South Africa); and its vice chairman, Marjorie Dart (South Africa).

The Temple Fay Institute of Academics and its director, Dr. Neil Harvey.

The Better Baby Institute and its director, Robert Derr; and staff members: Robert Derr, Jr. (United States), Jo Sherwood (Britain) and its qualified professional parents.

The Institutes for the Achievement of Human Potential (the parent organization) and its medical director, Robert Doman, M.D.; associate medical director, Roselise Wilkinson, M.D.; director in charge of children's affairs, Elaine Lee; and department heads: Helen Derr (United States), Lee Pattinson (Australia) and Bertha White, Rosalie Sherman, John Tini and Nancy McGuire (all of the United States).

Richard Norton, Senior Scientific Consultant.

The International Board of Directors: Walter J. Burke; Frank B. Cliffe; Rosa Collodel; Patrick Coyne; Robert Doman; James Dunn; Joseph Gay; Harry Guenther; Gretchen Kerr; Paul Laiolo; Frank D. McCormick; Kaname Matsuzawa; Samuel Metzger, III; Liza Minnelli; Clarence Mitchell; Robert Morris; Ralph Pelligra; Alessandro Pitturazzi; Marsden Proctor; Jose Carlos Veras; Chatham R. Wheat, III; and John Winthrop Wright.

Neither this book nor any of the other books the staff have written, which have so remarkably changed the lives of children the world over, would have been possible without the constant and generous support of the following people: John and Mary McShain, Louise Sacchi, Lloyd and Ellen Wells, Liza Minnelli, Samuel and Joan Metzger, Mr. and Mrs. C. B. Norris, Albert and Christine Vollmer, John and Josie Connelly, Mrs. Edward Cassard, Mrs. Henry Hoyt, Dan and Margaret Melcher, Walker G. Buckner, Gene and Lottie Losasso, Masaru and Yoshiko Ibuka, the hundreds of friends of the Institutes, the Sony Corporation and the hundreds of thousands of members of the United Steelworkers of America.

Finally, I wish to bow to all those in history who have believed with a consuming passion that children were really quite superior to the image that we adults have always held of them.

Those desiring more direct information may contact:

Glenn Doman, Director
The Institutes for the Achievement of
    Human Potential
8801 Stenton Avenue
Philadelphia, Pennsylvania 19118
U.S.A.